方芍堯 編著

新手入廚系列

巧手糕點

前言

廚房可以讓人忘卻煩憂，引發出浪漫因子，幻化成一件件令人垂涎的點心。特別是蒸好一盆盆晶瑩剔透、七彩繽紛、甜不膩口的糕點，佈滿一桌的時候，每一件糕點彷彿延續着我的夢想，為枯燥乏味的都市生活平添一點點生氣；所以我愛做糕點，它能再次煥發我的精神，恢復對生活的熱誠，向未來再次衝刺。

柔軟幼滑的糯米糕點、玲瓏的涼糕、香熱的煎餅……糕點是與人分享喜樂的好工具，蘿蔔糕、馬蹄糕等人人都吃過的糕點，加入不同的材料，糅合了點點情感，改變一點兒份量調配，口味又會煥然一新，奇味無窮。

朋友，不妨如我一樣走到廚房弄一些美味糕點，然後與自己的摯愛親朋分享，心情會變得很好，這是一個開心又快樂的經驗分享，試試吧！

目錄

啫喱果凍類

糯米糍丸子類

西式甜品類

日式甜品類

港式小食類

主料 Main Ingredients

馬蹄粉
water chestnut powder

粘米粉
rice flour

白玉粉
Japanese glutinous rice flour

糯米粉
glutinous rice flour

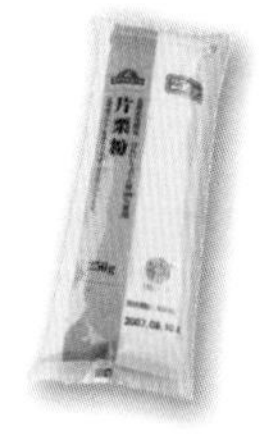

日本片栗粉
Japanese corn flour

蒸包粉
steamed bread mix

蒟蒻粉
konjac powder

艾草粉
ground Artemisia princeps

綠茶粉
green tea powder

副料 Sub-ingredientss

日本白芝麻
Japanese white sesame

日本黑芝麻
Japanese black sesame

日本有機黃豆粉
Japanese organic soybean powder

椰絲
desiccated coconut

西米
sago

鮮奶
fresh milk

紅腰豆
red kidney bean

黃桃
peach

雜豆
mixed beans

椰汁
coconut milk

腰果
cashew

馬蹄
water chestnut

核桃
walnut

藍莓
blue berry

草莓 / 士多啤梨
strawberry

石榴
pomegranate

芒果
mango

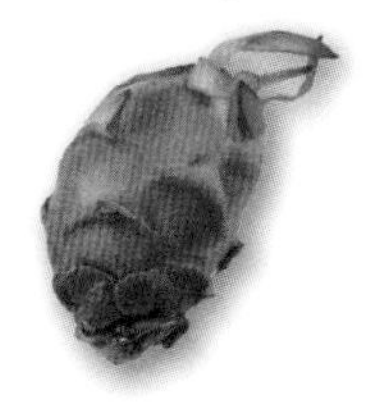

火龍果
dragon fruit

椰青
young coconut

6~8 人
Serves 6~8

40~50 分鐘
40~50 minutes

蒸雞蛋糕
Steamed Cake

材料 | Ingredients

雞蛋 8 隻	8 eggs
糖 280 克	280g sugar
麵粉 240 克	240g flour

做法 | Method

1. 雞蛋置於碗中，拂打至起泡。
2. 徐徐加入糖，攪拌至糖融解，蛋漿轉至淺黃色。
3. 麵粉篩勻，倒入（2）中，徐徐攪拌至粉和蛋漿混和成幼滑狀，靜置 20 分鐘。
4. 倒入蛋糕紙模上，用猛火隔水蒸約 20-30 分鐘。

1. Break eggs into a mixing bowl, then whisk until bubbles appear.
2. Add in sugar gradually, stir until sugar is dissolved and egg batter turns into light yellow.
3. Sift flour, pour into the mixture from step (2), fold into egg mixture slowly until smooth, then leave it for 20 minutes.
4. Pour into a cake mould and steam over high heat for about 20-30 minutes.

入廚貼士 | Cooking Tips

- 蒸糕時間要視乎糕盤大小來增減。
- Adjust cooking time in accordance with the size of cake mould.

馬拉糕

Brownish Sponge Cake

4~6 人
Serves 4~6

60~90 分鐘
60~90 minutes

材料 | Ingredients

雞蛋 6 隻	6 eggs
低筋麵粉 250 克	250g low gluten flour
紅糖 200 克	200g brown sugar
豬油 150 克	150g lard
泡打粉（發粉）1 湯匙	1 tbsp baking powder
梳打粉 1 茶匙	1 tsp soda powder

做法 | Method

1. 低筋麵粉、泡打粉與梳打粉一同篩勻。
2. 雞蛋放入大碗中，加入紅糖充份拂打至變淡白。
3. 分次加入豬油，充份攪拌至全部混和。
4. 將已篩的粉拌入（3）中，拌勻。
5. 注入糕盆內，放入蒸籠，以大火蒸 40 分鐘至熟透。

1. Sift low gluten flour, baking powder and soda powder together.
2. Break eggs into a mixing bowl, add in brown sugar and whisk thoroughly until milky white.
3. Add in lard gradually, stir thoroughly until fully blended.
4. Fold sifted flour into egg mixture from step (3) and mix thoroughly.
5. Pour into a cake mould and steam in a steamer over high heat for 40 minutes until cooked.

3~4 人
Serves 3~4

20~30 分鐘
20~30 minutes

栗子蒸糕
Steamed Chestnut Cake

材料 | Ingredients

即食栗子 5 粒	5 ready-to-eat chestnuts
雞蛋 3 隻	3 eggs
麵粉 100 克	100g flour
栗子蓉 50 克	50g chestnut purée
糖 50 克	50g sugar
泡打粉（發粉）1 茶匙	1 tsp baking powder

做法 | Method

1. 即食栗子切粒。
2. 雞蛋置於碗中與糖打至淡黃色。
3. 加入栗子茸拌勻，再加入已篩好的麵粉和泡打粉。
4. 倒入以掃油的蒸盤，隔水蒸 20-25 分鐘。

1. Dice ready-to-eat chestnuts.
2. Break eggs into a mixing bowl, whisk with sugar until light yellow.
3. Add in chestnut purée and mix well. Add in sifted flour and baking powder.
4. Brush some oil on a steaming tray, pour in batter and steam for 20-25 minutes.

三色蒸糕

Tri-colour Sponge Cake

2~3 人
Serves 2~3

30 分鐘
30 minutes

材料 | Ingredients

雞蛋 1 隻	1 egg
麵粉 150 克	150g flour
糖 50 克	50g sugar
牛奶 1/2 杯	1/2 cup milk
菜油 1 湯匙	1 tbsp vegetable oil
泡打粉（發粉）1 茶匙	1 tsp baking powder
可可粉、綠茶粉各少許	Pinch of cocoa powder and green tea powder

做法 | Method

1. 雞蛋打散後加入糖，用打蛋器充份攪拌，變成奶白色。
2. 慢慢加入菜油，攪拌至完全融合，再加入牛奶輕輕混合。
3. 麵粉和泡打粉一同篩勻後，加入蛋糊中拌勻。
4. 可將麵糊分成三份，其中兩份分別加入可可粉和綠茶粉混和。
5. 把三色粉糊分別置於小缽中，用大火蒸 15-20 分鐘即成。

1. Beat egg in a mixing bowl, add sugar and whip with an electric mixer until milky white.
2. Add vegetable oil slowly and stir until well mixed. Add milk and mix gently.
3. Sift flour and baking powder together, fold mixture into egg batter thoroughly.
4. Divide batter into 3 portions and mix 2 of them with cocoa powder and green tea powder respectively.
5. Place 3 portions of batter into 3 different bowls separately, steam over high heat for 15-20 minutes.

3~4 人
Serves 3~4

35~40 分鐘
35~40 minutes

馬蹄糕
Water Chestnut Pudding

材料 | Ingredients

馬蹄 150 克	150g water chestnut
紅糖 140 克	140g brown sugar
馬蹄粉 110 克	110g water chestnut powder
豬油 / 菜油 1/2 湯匙	1/2 tbsp lard / vegetable oil
清水 460 毫升	460 ml water

做法 | Method

1. 馬蹄洗淨，去皮，切薄片。
2. 燒熱鍋，加入紅糖與 360 毫升清水煮滾，再加入油和馬蹄煮 3-5 分鐘，熄火。
3. 用 100 毫升清水與馬蹄粉開勻成為馬蹄粉，將馬蹄粉漿倒入（2）中拌勻。
4. 倒入已掃油的糕盆上，用中大火隔水蒸 30 分鐘。

1. Rinse and peel water chestnuts and cut into thin slices.
2. Heat wok and cook brown sugar in 360ml of water and bring to a boil. Add oil and water chestnuts and cook for 3-5 minutes. Turn off heat.
3. Mix water chestnut powder with 100ml of water, pour into water chestnut mixture from step (2) and mix well.
4. Brush some oil in a mould, pour in batter and steam over medium to high heat for 30 minutes.

Turnip Cake with Chinese Sausages

臘味蘿蔔糕

4~6 人
Serves 4~6

1 小時
1 hour

材料 | Ingredients

臘腸 2 條	2 Chinese sausages
瑤柱 4 粒	4 dried scallops
蝦米 10 克	10g dried shrimps
白蘿蔔 800 克	800g turnip
粘米粉 400 克	400g rice flour
粟粉 400 克	400g corn flour
水 100 毫升	100ml water

調味料 | Seasonings

鹽 1 1/2 茶匙
胡椒粉適量

1 1/2 tsp salt
Pinch of pepper

做法 | Method

1. 材料洗淨。瑤柱用 100 毫升水浸透，撕碎，瑤柱水備用；蝦米浸軟，切粒。
2. 臘腸隔水蒸 10 分鐘，待涼後切片。
3. 白蘿蔔去皮，切絲；粘米粉和粟粉一同篩勻，用水開成粉漿。
4. 燒熱 1 湯匙油，爆香蝦米和臘腸，盛起。
5. 原鑊兜炒白蘿蔔，注入瑤柱水和調味料同煮 5-10 分鐘，熄火。
6. 將粉漿倒入（5）中，攪拌均勻，加入臘腸、蝦米和瑤柱拌勻，倒入已掃油的蒸盆，隔水蒸 45 分鐘。

1. Rinse ingredients. Soak dried scallops in 100ml of water until soft, tear into shreds and retain scallop water for later use. Soak dried shrimps until soft, cut into fine pieces.
2. Heat a wok and steam Chinese sausages for 10 minutes, leave to cool and cut into slices.
3. Peel turnips and cut into shreds. Sift rice flour and corn flour together, mix with water into batter.
4. Heat 1 tbsp of oil, sauté dried shrimps and Chinese sausages until fragrant, take out and set aside.
5. Put turnips into the same wok, add scallop water and seasonings and cook for 5~10 minutes, turn off heat.
6. Pour batter into turnip mixture from step (5), stir thoroughly, add sausages, dried shrimps and scallops and mix well. Brush some oil in a cake mould, pour in batter and steam for 45 minutes.

4~6 人
Serves 4~6

25~30 分鐘
25~30 minutes

紅棗人參糕

Red Dates and Ginseng Cake

材料 | Ingredients

鮮紅棗 2 粒	2 fresh red dates
鮮人參 1 茶匙（切碎）	1 tsp fresh ginseng (finely chopped)
雞蛋 4 隻	4 eggs
低筋麵粉 100 克	100g low gluten flour
糖 60 克	60g sugar
牛奶 50 毫升	50ml milk
泡打粉（發粉）1/2 茶匙	1/2 tsp baking powder

做法 | Method

1. 鮮紅棗和鮮人參分別切碎備用。
2. 雞蛋至於碗中打散，加入糖拂打至淺黃色。
3. 加入牛奶後，再加入已篩的低筋麵粉和泡打粉，拌勻。
4. 再將紅棗碎和人參碎拌入，倒入模中，大火隔水蒸 15-20 分鐘。

1. Chop up fresh red dates and ginseng separately, set aside.
2. Beat eggs in a mixing bowl, add sugar and whisk until light yellow.
3. Add milk, sifted low gluten flour and baking powder and stir thoroughly.
4. Fold in chopped red dates and ginseng, pour mixture into a mould, steam over high heat for 15-20 minutes.

綠茶酥餅

Green Tea Pastry

8 個
8 pcs

25~30 分鐘
25~30 minutes

粉糰 | Dough

雞蛋 1/2 隻　1/2 egg
低筋麵粉 110 克　110g low gluten flour
牛油 75 克　75g butter
糖 20 克　20g sugar
奶粉 15 克　15g milk powder

餡料 | Filling

白豆蓉 200 克
綠茶粉 2 茶匙
綠茶香醬 1/2 茶匙

200g white bean purée
2 tsps green tea powder
1/2 tsp green tea sauce

做法 | Method

1. 牛油置室溫放軟，放大碗中，加入糖一起攪至淡黃色。
2. 雞蛋打散，分次加入（1）內。奶粉和低筋麵粉篩匀，加入蛋漿中，搓成粉糰。
3. 白豆蓉加入綠茶粉和綠茶香醬，搓匀成為綠茶餡料。
4. 將粉糰分為 2 份，稍為搓長後，包入 1/2 份綠茶餡，收口，搓成將 10 厘米長，切成 4 份，剩餘材料如上做法。
5. 放入已預熱的焗爐，用 170℃焗 15 分鐘，取出，掃上少許蛋漿，再焗 3 分鐘，即成。

1. Place butter in room temperature until soft. Add in sugar and stir until light yellow in a bowl.
2. Beat egg and add into butter and sugar mixture from step (1) gradually. Sift milk powder and low gluten flour, add to egg mixture and knead into a dough.
3. Add green tea powder and green tea sauce into green bean purée, knead well into green tea filling.
4. Divide the dough into 2 portions. Knead a portion into an oblong shape, wrap in half of green tea filling, seal well. Knead into a rod of 10 cm long and cut it into 4 pieces. Repeat this process with remaining ingredients.
5. Place pastries onto a preheated oven and bake at 170℃ for 15 minutes. Take out and brush some egg liquid on surface. Bake for another 3 minutes and serve.

19~20 粒
19~20 pcs

25~30 分鐘
25~30 minutes

奶黃水晶餅
Egg Custard Cake

粉糰 | Dough

澄麵 50 克	50g tang starch
糖 36 克	36g sugar
生粉 12 克	12g cornstarch
豬油 1 茶匙	1 tsp lard
水 125 毫升	125ml water

奶黃餡 | Egg Custard Filling

蛋黃 2 隻	2 egg yolks
日本片粟粉 60 克	60g Japanese corn flour
糖 40 克	40g sugar
牛油溶液 30 克	30g melted butter
吉士粉 30 克	30g custard powder
奶粉 15 克	15g milk powder
煉奶 2 湯匙	2 tbsps condensed milk
椰漿 2 湯匙	2 tbsps coconut milk

做法 | Method

1. 將奶黃餡料全部放在碗內，隔水邊煮邊拌勻，直至奶黃糊濃稠，取出待涼，分成每份約重 20 克，搓成球狀。
2. 水煮滾，加入糖和豬油煮融，撞入已篩的澄麵和生粉中，拌至透明。
3. 將麵糰分割成每個 25 克，用麵桿推薄麵糰，包上奶黃餡，印入餅模。
4. 用大火隔水蒸 10-15 分鐘，掃油即成。

1. Place all egg custard filling ingredients into a mixing bowl, put above hot water, stir continuously until thick. Take out and leave to cool. Divide into portions of about 20g each and knead into balls.
2. Bring water to a boil, add sugar and lard until dissolved. Pour into sifted tang starch and cornstarch and stir until translucent.
3. Divide dough into portions of 25g each. Roll out into a thin slice, wrap in egg custard filling and stuff into a cake mould.
4. Steam cakes over high heat for 10-15 minutes, brush some oil on top and serve.

荔蓉紫菜糕

Taro and Seaweed Cake

3~4 人
Serves 3~4

30~35 分鐘
30~35 minutes

材料 | Ingredients

荔芋 200 克	200g taro
粟粉 1 茶匙	1 tsp cornstarch
鹽 1/4 茶匙	1/4 tsp salt
紫菜適量	Some seaweed
胡椒粉適量	Pinch of pepper

做法 | Method

1. 荔芋洗淨，去皮，切角，隔水蒸 20-25 分鐘。
2. 荔芋用匙壓成蓉，加入鹽、粟粉和胡椒粉拌勻。
3. 用手搓成三角形，再在外面沾上一片紫菜。
4. 燒熱鍋，用慢火煎荔芋至兩面金黃色，即可享用。

1. Rinse and peel taro, cut into wedges. Heat a wok and steam for 20-25 minutes.
2. Mash taro into purée, add salt, cornstarch and pepper and stir thoroughly.
3. Knead taro with hands into triangular shape, stick a piece of seaweed on it.
4. Heat a wok and pan-fry taro over low heat until both sides turn brownish. Ready to serve.

健康三色豆糕
Assorted Bean Cake

紅腰豆糕 | Red Kidney Bean Cake

罐裝紅腰豆適量	Some canned red kidney beans
冰糖 15 克，水 150 毫升	15g rock sugar, 150ml water
魚膠粉 5 克，水 25 毫升	5g gelatine powder, 25 ml water

綠豆糕 | Green Bean Cake

開邊綠豆 80 克
綠茶香油少許
冰糖 20 克，水 400 毫升
魚膠粉 5 克，水 25 毫升

80g green beans (cut into halves)
some green tea essence
20g rock sugar, 400ml water
5g gelatine powder, 25 ml water

眉豆糕 | Eyebrow Bean Caked

眉豆 100 克	100g eyebrow beans
冰糖 20 克，水 400 毫升	20g rock sugar, 400ml water
魚膠粉 5 克，水 25 毫升	5g gelatine powder, 25ml wate

做法 | Method

1. 紅腰豆糕：魚膠粉置於碗中水拌勻，然後座於熱水中拌至清澈。燒熱鍋，糖與水煮融，加入魚膠溶液拌勻。待微涼後，拌入紅腰豆，倒入糕模中雪至凝固。
2. 綠豆糕：開邊綠豆用清水浸 20 分鐘。燒熱鍋，水煮滾後，加入綠豆煮至微開，再加入冰糖，熄火，加入魚膠溶液和綠茶香油，待涼至微凝稠才倒入已雪至凝固的紅腰豆糕上。
3. 眉豆糕：眉豆用清水浸 20 分鐘，燒熱鍋，水煮滾後加入眉豆煮至微開，再加入冰糖，熄火，再加入魚膠溶液，待涼至微稠才倒入已雪至凝固的綠豆糕上。
4. 待凝固後，切件享用。

1. Red Kidney Bean Cake: Mix gelatine powder with water in a bowl and melt above hot water until the liquid becomes clear. Heat a wok and boil sugar with water until dissolved, add gelatine liquid and mix well. When it is cool, mix with red kidney beans, pour into a mould and refrigerate until set.
2. Green Bean Cake: soak green beans in water for 20 minutes. Bring water to a boil, add green beans and cook until open up. Add rock sugar, turn off heat, add in gelatine liquid and green tea essence. Leave to cool until it starts to solidify. Pour mixture on top of the solidified red kidney bean cake.
3. Eyebrow Bean Cake: Soak eyebrow beans in water for 20 minutes. Bring water to a boil, add eyebrow beans and cook until open up. Add rock sugar. Turn off heat, add gelatine liquid, leave mixture to cool down until it starts to solidify. Pour onto solidified green bean cake.
4. Leave to solidify, cut into pieces and serve.

南瓜糕 Pumpkin Cake

4~6 人
Serves 4~6

30~40 分鐘
30~40 minutes

材料 | Ingredients

日本南瓜 200 克	200g Japanese pumpkin
淡忌廉 450 克	450g whipping cream
椰汁 100 毫升	100ml coconut milk
糖 40 克	40g sugar
粟粉 20 克	20g cornstarch
魚膠粉 8 克	8g gelatine powder
水 150 毫升	150ml water

做法 | Method

1. 魚膠粉置於碗中，用水 50 毫升拌勻，座於熱水中拌至清澈。
2. 南瓜去皮，切片。燒熱鍋，隔水蒸 15~20 分鐘至軟腍，待涼後壓成蓉。
3. 燒熱鍋，加入椰汁、淡忌廉、糖和水 100 毫升煮滾，加入南瓜蓉，再用少許水開勻粟粉，邊煮邊拌勻，煮滾後熄火，再加入魚膠溶液拌勻。
4. 待涼後放入雪櫃中冷藏至凝固，食用時切件即可。

1. Put gelatine powder into a bowl and mix with 50 ml of water, put the bowl above hot water and stir until liquid becomes clear.
2. Peel pumpkin, cut into slices. Heat a wok and steam for 15~20 minutes until soft. Leave to cool and mash into purée.
3. Put coconut milk, whipping cream, sugar and 100 ml of water in a pot and bring to a boil. Add pumpkin purée, then stir in cornstarch and water mixture while cooking. Bring to a boil, turn off heat, then add in gelatine liquid and stir thoroughly. Leave to cool.
4. Set in a refrigerator, cut it into slices and serve.

4~6 人
Serves 4~6

35~45 分鐘
35~45 minutes

鴛鴦豆糕

Mixed Bean Pudding

材料 | Ingredients

綠豆 60 克	60g green bean
罐裝紅腰豆 4 湯匙	4 tbsps canned kidney bean
椰汁 100 毫升	100ml coconut milk
糖 30 克	30g sugar
粟粉 20 克	20g cornstarch
魚膠粉 8 克	8g gelatine powder
綠茶香醬 1/2 湯匙	1/2 tbsp green tea sauce
水 1040 毫升	1040ml water

做法 | Method

1. 綠豆浸透，置鍋內，注入清水 1000 毫升，煮沸後轉中火煮至軟脸及綠豆水約剩 300 毫升。
2. 加入糖和椰汁煮滾後，再拌入粟粉水（粟粉用少許水開勻），拌勻，熄火。
3. 魚膠粉置碗中與水 40 毫升拌勻，座於熱水中拌至清澈。
4. 將魚膠粉溶液拌入（2），加入綠茶香醬和紅腰豆拌勻，倒入模中，放入雪櫃中冷藏至凝固。
5. 食用時，可伴以淡奶和士多啤梨蓉。

1. Soak green beans in water, place in a pot, fill in 1000ml of water and bring to a boil. Cook over medium heat until tender with liquid reduced to about 300ml.
2. Add sugar and coconut milk and bring to a boil. Stir in cornstarch liquid (mix cornstarch with some water), mix well and turn off heat.
3. Mix gelatine powder with 40 ml of water in a bowl, melt it above hot water until the liquid becomes clear.
4. Add gelatine liquid into coconut milk mixture from step (2). Add green tea sauce and kidney beans, mix thoroughly, pour mixture into a mould and set in refrigerator.
5. Serve with evaporated milk and strawberry purée.

蓮子桂花糕

Lotus Seed and Osmanthus Fragrans Pudding

3~4 人
Serves 3~4

40~45 分鐘
40~45 minutes

材料 | Ingredients

蓮子 30 粒	30 lotus seeds
冰糖 30 克	30g rock sugar
桂花糖 2 茶匙	2 tsps osmanthus fragrans sugar
大菜 2 克	2g agar agar
水 320 毫升	320 ml water

做法 | Method

1. 大菜用水浸 2~3 小時。
2. 燒熱鍋，蓮子用清水煮 30 分鐘，加入冰糖和大菜煮融。
3. 待呈微凝稠，再加入桂花糖，輕輕拌勻，倒入模中，放入雪櫃中冷藏至凝固。

1. Soak agar agar in water for 2~3 hours.
2. Heat wok, cook lotus seeds in water for 30 minutes, add rock sugar and agar agar, cook until dissolved.
3. Leave mixture until it starts to thicken. Add osmanthus fragrans sugar and stir slightly. Pour into a mould and set in refrigerator.

4~6 人
Serves 4~6

20 分鐘
20 minutes

綠茶西米糕
Green Tea and Sago Pudding

材料 | Ingredients

牛奶 100 毫升	100ml milk
淡忌廉 100 毫升	100ml whipping cream
糖 40 克	40g sugar
西米 4 湯匙	4 tbsps sago
魚膠粉 8 克	8g gelatine powder
綠茶粉 5 克	5g green tea powder
水 40 毫升	40ml water

做法 | Method

1. 魚膠粉置於碗中，與水拌勻，座於熱水中拌至清澈。
2. 西米用水浸 30~40 分鐘。燒熱鍋，用水煮透後過冷河。淡忌廉於碗中拂打至軟身（濕性軟化）。
3. 燒熱鍋，加入牛奶、糖和綠茶粉煮 5~8 分鐘，過篩，加入魚膠溶液拌勻。
4. 待綠茶奶漿冷卻至微凝稠時，才加入已拂打的淡忌廉和西米拌勻，倒入糕模中，放入雪櫃中冷藏雪至凝固。
5. 食用時，切件，可拌入淡奶一同享用。

1. Place gelatine powder in a bowl, mix with water, then melt above hot water until the liquid becomes clear.
2. Soak sago in water for 30~40 minutes, Heat wok and boil with water until cooked, then rinse with cold water. Whisk whipping cream until soft peak.
3. Place milk, sugar and green tea powder in a pot and cook for 5~8 minutes, strain and mix with gelatine liquid.
4. Leave green tea mixture to cool until thickened. Add whipped cream and sago and mix well. Pour mixture into a mould and set in refrigerator.
5. Cut pudding into pieces and serve with evaporated milk.

雪耳杞子糕

White Fungus and Lycium Berry Pudding

3~4 人
Serves 3~4

15~20 分鐘
15~20 minutes

材料 | Ingredients

杞子 30 粒	30 lycium berries
雪耳 2 盞	2 bunches white fungus
冰糖 30 克	30g rock sugar
魚膠粉 8 克	8g gelatine powder
水 390 毫升	390ml water

做法 | Method

1. 雪耳用水浸發開，剪去底部硬梗。
2. 魚膠粉置於碗中，與 40 毫升水拌勻，座於熱水中拌至清澈。
3. 燒熱鍋，加入 350 毫升水煮滾，放入杞子和雪耳，用中火煮 15~20 分鐘。
4. 加入冰糖煮融，熄火。再加入已清澈的魚膠溶液拌勻。
5. 倒入糕模中，放入雪櫃中冷藏雪至凝固。

1. Soak white fungus in water until expanded. Cut off hard portion at the bottom.
2. Place gelatine powder in a bowl, mix with 40ml of water, then melt above hot water until liquid becomes clear.
3. Bring 350ml of water to a boil, add lycium berries and white fungus and cook over medium heat for 15~20 minutes.
4. Add rock sugar and cook until dissolved. Turn off heat. Add clear gelatine liquid and mix well.
5. Pour mixture into a mould, set in refrigerator. Serve.

4~6 人
Serves 4~6

10~20 分鐘
10~20 minutes

香滑椰汁糕

Coconut Pudding

材料 | Ingredients

椰汁 400 毫升	400ml coconut milk
淡忌廉 240 克	240g whipping cream
糖 180 克	180g sugar
魚膠粉 20 克	20g gelatine powder
蛋白 4 隻	4 egg whites
凍水 100 毫升	100ml cold water

做法 | Method

1. 魚膠粉盛於碗中，加入凍水開勻，座於熱水中拌至清澈，備用。
2. 椰汁、淡忌廉和 100 克糖煮滾後，加入魚膠溶液，拌勻待涼。
3. 蛋白於碗中打起後，分次加入糖，打至濃稠，拌入已涼凍的椰奶漿中拌勻，倒入糕模中，放入雪櫃冷藏至凝固。
4. 食用時可切成不同形狀。

1. Place gelatine powder in a bowl, stir in cold water and melt above hot water until liquid becomes clear. Set aside.
2. Place coconut milk, whipping cream and 100g of sugar in a pot and bring to a boil. Add gelatine liquid and stir well, leave to cool.
3. Whisk egg whites until firm peak, add sugar gradually and whisk until thick. Stir in cool coconut milk sauce. Pour mixture into a mould and set in refrigerator.
4. Cut pudding into different shapes and serve.

4~6 人
Serves 4~6

40~45 分鐘
40~45 minutes

紅豆椰汁糕

Red Bean Coconut Cake

材料 | Ingredients

椰汁 240 克	240g coconut milk
鮮奶 120 毫升	120 ml milk
糖 80 克	80g sugar
紅豆 50 克	50g red beans
粟粉 20 克	20g cornstarch
魚膠粉 10 克	10g gelatine powder
水 150 毫升	150ml water

飾面 | Garnishing

椰絲適量
Some desiccated coconut

做法 | Method

1. 紅豆用水浸 30 分鐘。燒熱鍋，加水煮至腍軟，盛起。
2. 魚膠粉置碗中，用 50 毫升水拌勻，座於熱水中拌至清澈。
3. 將椰汁、鮮奶、糖、粟粉和 100 毫升水同置碗中拌勻。燒熱鍋，用慢火（邊煮邊拌勻）煮滾，熄火。
4. 加入魚膠溶液拌勻，再加入紅豆拌勻，倒入糕模中，置於雪櫃中冷藏至凝固。
5. 食用時，取出切件，沾上椰絲即可。

1. Soak red beans in water for 30 minutes. Heat a wok, add some water and boil red beans until soft, then set aside.
2. Place gelatine powder in a bowl, mix with 50ml of water, then melt it above hot water until liquid becomes clear.
3. Place coconut milk, milk, sugar, cornstarch and water in a bowl and mix well. Cook it over low heat, stir until it boils and turn off heat.
4. Add gelatine powder, red beans and mix well. Pour into a cake mould and set in refrigerator.
5. Cut into slices, coat with desiccated coconut and serve.

入廚貼士 | Cooking Tips

- 紅豆 50 克亦可用即食紅豆約 4 湯匙代替。
- 50g red beans can be replaced by 4 tbsps of ready-to-eat red beans instead.

繽紛鮮奶糕

Colourful Milk Pudding

鮮奶糕 | Milk Pudding

鮮奶 400 毫升
淡忌廉 150 克（拂起）
糖 50 克
魚膠粉 10 克
雲呢拿香油 1/4 茶匙
水 50 毫升

400g milk
150g whipping cream (whisked)
50g sugar
10g gelatine powder
1/4 tsp vanilla essence
50 ml water

4~6 人
Serves 4~6

10~15 分鐘
10~15 minutes

彩色啫喱粒 | Colourful Jelly Cubes

糖 40 克	40g sugar
魚膠粉 15 克	15g gelatine powder
檸檬皮青 1 個	1 lemon zest
檸檬 1/2 個（榨汁）	1/2 lemon (squeezed juice)
食用色素少許（任何顏色）	A little edible artificial colourings (any colours)
熱水 400 毫升	400ml hot water

做法 | Method

1. 繽紛啫喱粒：魚膠粉加入糖後，與熱水拌勻，再加入檸檬汁和檸檬皮青，分別倒入器皿中，加入不同顏色的色素拌勻。放入雪櫃冷藏至凝固，切粒。
2. 鮮奶糕：魚膠粉與水拌勻。鮮奶加糖煮至約 80℃，熄火，加入魚膠溶液拌勻，拌入淡忌廉，再加入雲呢拿香油拌勻。
3. 組合：將奶糊倒入器皿中，待開始凝固時，放上繽紛啫喱粒，放回雪櫃冷藏至凝固，切件即成。

1. Colourful Jelly Cubes: Add sugar into gelatine powder and mix with hot water. Add lemon juice and lemon zest, pour mixture into several containers and stir in different colourings respectively. Place into refrigerator until liquid solidifies into jelly. Dice.
2. Milk Pudding: Mix gelatine powder with water. Add sugar into milk and cook to 80℃ . Turn off heat, add gelatine liquid and stir thoroughly. Stir in whipping cream and add vanilla essence and mix well.
3. Combination: Pour milk batter into a container and leave until it starts to solidify. Place colourful jelly cubes on top and set in refrigerator. Cut into pieces and serve.

4~6 人
Serves 4~6

45~60 分鐘
45~60 minutes

西米芒果凍糕

Sago and Mango Chilled Pudding

材料 | Ingredients

芒果 1 個（切粒）	1 mango (diced)
西米 6 湯匙	6 tbsps sago
糖 60 克	60g sugar
蒟蒻粉 15 克	15g konjac powder
檸檬青 1 個	1 lemon zest
檸檬 1/2 個（榨汁）	1/2 lemon (squeezed juice)
熱水 300 毫升	300ml hot water

做法 | Method

1. 西米用水浸 30~40 分鐘。燒熱鍋，用水煮透後過冷河，瀝乾。
2. 蒟蒻粉和糖放碗中拌勻，加入熱水拌勻至清徹。
3. 拌入檸檬青和檸檬汁。
4. 加入西米和芒果粒拌勻，倒入模中，放進雪櫃冷藏至凝固。

1. Soak sago in water for 30~40 minutes. Heat a wok and cook with water until done, rinse with cold water, drain.
2. Mix konjac powder with sugar. Add hot water and stir well until liquid becomes clear.
3. Stir in lemon zest and lemon juice.
4. Add sago and mango dice and mix well. Pour mixture into a mould and set in refrigerator. Ready to serve.

士多啤梨牛奶凍
Strawberry Milk Chilled Pudding

3~4 人
Serves 3~4

40 分鐘
40 minutes

牛奶凍 | Milk Pudding

牛奶 400 毫升
糖 25 克
大菜 5 克
水 150 毫升

400ml milk
25g sugar
5g agar agar
150ml water

糖漿 | Syrup

糖 25 克
檸檬青 1/2 個
士多啤梨適量
水 250 毫升

25g sugar
1/2 lemon zest
Some strawberries
250ml water

做法 | Method

1. 大菜用水浸 2~3 小時。燒熱鍋，加入清水煮融，再加入牛奶和糖同煮融，過篩。
2. 倒入糕模中，放進雪櫃冷藏至凝固，取出，切粒。
3. 燒熱鍋，加入糖和水煮成糖漿，待涼。
4. 士多啤梨切粒，再與糖漿和牛奶凍拌勻，即可享用。

1. Soak agar agar in water for 2~3 hours. Heat a wok, add water and cook until dissolved. Add milk and sugar and cook until dissolved, strain.
2. Pour mixture into a mould, then set in refrigerator. Take out and dice.
3. Heat a wok, cook sugar and water into a syrup, then leave to cool.
4. Dice strawberries, then mix with syrup and milk pudding. Ready to serve.

4~6 人
Serves 4~6

10~15 分鐘
10~15 minutes

Sesame Seed Pudding
芝麻布甸

材料 | Ingredients

淡忌廉 150 克	150g whipping cream
芝麻粉 100 克	100g sesame seed powder
糖 60 克	60g sugar
魚膠粉 8 克	8g gelatine powder
鮮奶 250 毫升	250ml milk
水 40 毫升	40 ml water

做法 | Method

1. 魚膠粉置於碗中，加入清水拌勻，座於熱水中拌至清澈。
2. 鮮奶、糖和芝麻粉一同放入鍋中，加熱至 80℃，熄火，加入魚膠溶液拌勻，待涼。
3. 淡忌廉拂打至軟身（即濕性軟化）後，拌入已微凝稠的奶糊中，拌勻。
4. 倒入糕模中，放入雪櫃冷藏至凝固即可。

1. Place gelatine powder in a bowl, stir in water and melt above hot water until liquid becomes clear.
2. Place milk, sugar and sesame seed powder in a pot, and boil it to 80℃ , turn off heat. Add gelatine liquid and mix well, leave to cool.
3. Whisk whipping cream until soft peak, stir in thickened milk sauce and mix well.
4. Pour mixture into a mould, set in refrigerator and serve.

綠茶布甸

Green Tea Pudding

4~6 人
Serves 4~6

10 分鐘
10 minutes

材料 | Ingredients

蛋黃 1 隻	1 egg yolk
淡忌廉 100 克	100g whipping cream
糖 60 克	60g sugar
綠茶粉 14 克	14g green tea powder
魚膠粉 10 克	10g gelatine powder
鮮奶 300 毫升	300 ml milk
水 60 毫升	60ml water

做法 | Method

1. 魚膠粉置於碗中，用水調勻，座於熱水中拌至清澈。
2. 鮮奶、糖、忌廉和綠茶粉放鍋中，加熱至 80℃，撞入已打散的蛋黃中，過篩。
3. 加入魚膠溶液，倒入模中，放入雪櫃冷藏至凝固。

1. Place gelatine powder in a bowl, add water and mix well. Melt above hot water until liquid becomes clear.
2. Put milk, sugar, cream and green tea powder in a pot and boil to 80℃ . Pour mixture into beaten egg yolk, strain.
3. Add gelatine liquid and pour into a mould, then set in refrigerator.

3~4 人
Serves 3~4

15 分鐘
15 minutes

芒果布甸
Mango Pudding

材料 | Ingredients

芒果 1 個	1 mango
蛋黃 1 隻（打散）	1 egg (beaten)
芒果蓉 200 克	200g mango purée
淡忌廉 100 克	100g whipping cream
糖 50 克	50g sugar
魚膠粉 10 克	10g gelatine powder
鮮奶 200 毫升	200 ml milk
水 60 毫升	60 ml water

做法 | Method

1. 將芒果切粒備用；魚膠粉置碗中加入清水拌匀，座於熱水中拌至清澈。
2. 淡忌廉、糖、鮮奶和半份芒果蓉放鍋中，加熱煮滚，熄火。再加入蛋黃、剩餘芒果蓉和魚膠溶液拌匀。
3. 將芒果粒拌入（2），倒入器皿中，放入雪櫃冷藏至凝固，食用時可拌入淡奶。

1. Dice mango and set aside. Put gelatine powder in a bowl, stir in water, melt above hot water until mixture becomes clear.
2. Put whipping cream, sugar, milk and half portion of mango purée in a pot, bring to a boil and turn off heat. Add egg yolk, remaining mango puree and gelatine mixture, mix thoroughly.
3. Add mango dice into pudding mixture from step (2), pour into a container and chill in refrigerator. Serve with evaporated milk.

入廚貼士 | Cooking Tips

- 芒果蓉可用鮮芒果攪碎或購買現貨果蓉。
- Mango purée can be made by beating fresh mango flesh into a paste or purchase from supermarket.

鮮果啫喱凍

Fresh Fruit Jelly

提子啫喱｜Grape Jelly

藍莓適量
濃縮提子（葡萄）汁 3 湯匙
糖 20 克、魚膠粉 5 克
熱水 100 毫升、水 25 毫升

Some blueberries
3 tbsps condensed grape juice
20g sugar, 5g gelatine powder
100ml hot water, 25 ml water

4~6 人
Serves 4~6

60~90 分鐘
60~90 minutes

士多啤梨啫喱｜Strawberry Jelly

士多啤梨適量（切粒） Some strawberries (diced)
檸檬 1/2 個（榨汁） 1/2 lemon (squeezed juice)
糖 35 克、魚膠粉 5 克 35g sugar, 5g gelatine powder
熱水 100 毫升、水 25 毫升 100ml hot water, 25 ml water

蜜桃啫喱｜Peach Jelly

罐裝水蜜桃適量（切粒） Some canned peaches (diced)
濃縮橙汁 3 湯匙 3 tbsps condensed orange juice
糖 20 克、魚膠粉 5 克 20g sugar, 5g gelatine powder
熱水 100 毫升、水 25 毫升 100ml hot water, 25 ml water

做法｜Method

1. 提子啫喱：魚膠粉置碗中，用水拌勻，座於熱水中拌至清澈。提子汁和糖用熱水置碗中拌勻，加入魚膠溶液拌勻，待涼，加入藍莓，倒入模中，放入雪櫃冷藏至凝固，做成第一層。
2. 士多啤梨啫喱：檸檬汁和糖置碗中，用水拌勻，加入魚膠溶液，拌勻，待涼，加入士多啤梨粒，倒進提子啫喱上，放入雪櫃冷藏至凝固。
3. 水蜜桃啫喱：橙汁和糖用熱水拌勻，加入魚膠溶液拌勻，待涼，加入水蜜桃粒，倒進啫喱上，放入雪櫃冷藏至凝固後用熱水敷數秒，倒模扣出。

1. Grape Jelly: place gelatine powder in a bowl and mix with water. Melt above hot water until liquid becomes clear. Mix grape juice with sugar and hot water. Add gelatine liquid and stir well, leave to cool. Add in blueberries, pour mixture into a mould, then set in refrigerator, the first layer is made.
2. Strawberry Jelly: mix lemon juice with sugar and hot water. Add gelatine liquid and stir well, leave to cool. Add strawberries, pour over grape jelly, then set in refrigerator.
3. Peach Jelly: mix orange juice with sugar and hot water. Add gelatine liquid and mix well, leave to cool. Add peach dice, pour on top of strawberry and grape jelly, then set in refrigerator. Place jelly mould above hot water for a few seconds, flip it over and jelly will slip out easily. Serve.

4~6 人
Serve 4~6

45~50 分鐘
45~50 minutes

香茅青檸凍
Chilled Pudding with Lemongrass and Lime

材料 | Ingredients

香茅 2 條	2 stalks lemongrass
青檸 1 個（榨汁）	1 lemon juice (squeezed juice)
檸檬皮青 1 個	1 lemon zest
冰糖 50 克	50 g rock sugar
魚膠粉 15 克	15 g gelatine powder
水 750 毫升	750 ml water
凍水 80 毫升	80 ml cold water

做法 | Method

1. 魚膠粉至碗中，加入凍水拌勻，座於熱水中拌至清澈，備用。
2. 香茅切碎。燒熱鍋，與水同煮 30 分鐘，加入冰糖煮至糖完全融解，熄火。加入魚膠溶液，青檸汁和檸青拌勻。
3. 倒入模中，放入雪櫃冷藏至凝固，切粒。可加少許檸青伴吃。

1. Put gelatine powder in a bowl, add cold water and mix well. Melt above hot water until liquid becomes clear, then set aside.
2. Chop up lemongrass finely and cook with water for 30 minutes. Add rock sugar and cook until completely dissolved. Turn off heat, add gelatine liquid, lemon juice and lemon zest and mix well.
3. Pour mixture into a mould and set in refrigerator. Dice pudding. Serve with a pinch of lemon zest.

鴛鴦咖啡凍

Yin-Yang Coffee Chilled Pudding

檸檬咖啡凍 | Lemon Coffee Pudding

即溶咖啡 1/2 茶匙，檸檬青 1/2 個
糖 40 克，魚膠粉 4 克
熱水 100 毫升，水 20 毫升

1/2 tsp instant coffee, 1/2 lemon zest
40g sugar, 4g gelatine powder
100 ml hot water, 20ml water

4~6 人
Serves 4~6

10~15 分鐘
10~15 minutes

牛奶咖啡凍｜Milk Coffee Pudding

即溶咖啡 1 茶匙，蛋黃 2 個
淡忌廉 100 克，糖 50 克
咖啡香醬少許
牛奶 50 毫升
魚膠粉 6 克，水 30 毫升

1 tsp instant coffee, 2 egg yolks,
100g whipping cream, 50g sugar
Some coffee sauce
50 ml milk
6g gelatine powder, 30ml water

做法｜Method

1. 檸檬咖啡凍：魚膠粉置於碗中，用 20 毫升水拌勻，座於熱水中拌至清澈。
2. 即溶咖啡、糖置於碗中，用熱水開溶，下魚膠溶液和檸檬青拌勻，待涼後倒入模中約三分滿，放入雪櫃冷藏至凝固。
3. 牛奶咖啡凍：蛋黃打散；即溶咖啡用 3 湯匙置於碗中，用熱水開溶；淡忌廉拂打好。
4. 燒熱鍋，加入牛奶和糖，加熱至 80℃，撞入蛋黃中，拌勻。
5. 加入咖啡溶液、魚膠溶液、淡忌廉和咖啡香醬拌勻，倒入檸檬咖啡凍上，放入雪櫃冷藏至凝固。

1. Lemon Coffee Pudding: Place gelatine powder in a bowl and mix with water, melt above hot water until liquid becomes clear.
2. Place instant coffee and sugar in a bowl and dissolve in hot water. Add gelatine liquid and lemon zest and mix well, leave to cool. Pour into a mould until it is 30% full and set in refrigerator.
3. Milk Coffee Pudding: Beat egg yolk. Dissolve instant coffee with 3 tbsps of hot water in a bowl. Whisk whipping cream until thick.
4. Heat a wok, add milk and sugar and heat to 80℃ . Pour into egg yolks and mix well.
5. Add coffee liquid, gelatine liquid, whipping cream and coffee paste, mix well. Pour over lemon coffee pudding and set in refrigerator.

4~6 人
Serves 4~6

5~10 分鐘
5~10 minutes

鮮果水晶凍
Fresh Fruit Jelly Pudding

材料 | Ingredients

檸檬青 1 個
檸檬 1/2 個（榨汁）
鮮果（藍莓、士多啤梨）適量
糖 60 克
蒟蒻粉 15 克
熱水 300 毫升

1 lemon zest
1/2 lemon (squeezed juice)
Some fresh fruit
(blueberries, strawberries)
60g sugar
15g konjac powder
300ml hot water

做法 | Method

1. 鮮果洗淨，用布吸乾水分。
2. 蒟蒻粉與糖放碗中拌勻，加入熱水拌勻。
3. 加入檸檬青和檸檬汁攪拌勻。
4. 倒入已放鮮果的啫喱杯中，冷藏至凝固即可享用。

1. Wash fresh fruit with water and then pat dry.
2. Mix konjac powder with sugar in a bowl, add hot water and stir thoroughly.
3. Add lemon zest and lemon juice and mix well.
4. Place some fresh fruit in several jelly cups, pour in lemon mixture, then set in refrigerator. Ready to serve.

擂沙湯丸

Coated Glutinous Rice Balls

20 粒
20 pcs

5~10 分鐘
5~10 minutes

粉糰 | Dough

糯米粉 150 克
粘米粉 20 克
豬油 1/2 湯匙
暖水 150 毫升

150g glutinous rice flour
20g rice flour
1/2 tbsp lard
150ml warm water

餡料 | Filling

麻蓉 180 克
180g sesame seed purée

飾面 | Garnish

日本黃豆粉適量
Some Japanese soy bean powder

做法 | Method

1. 糯米粉與粘米分放碗中篩勻，加入豬油和暖水搓成粉糰，搓長，分成 20 小份。
2. 麻蓉分成 20 小分，搓圓。
3. 將粉糰搓圓按扁，加入麻蓉，封口，搓圓，置大滾水中煮至湯丸浮起。
4. 撈起，瀝乾水分，沾上黃豆粉即成。

1. Sift glutinous rice flour and rice flour in a bowl, add lard and warm water. Knead into a long rod shape and divide into 20 portions.
2. Divide sesame seed purée into 20 portions and knead into balls.
3. Roll a dough portion into a ball and then press to flatten. Wrap in sesame purée, seal well and roll into a ball. Cook rice balls in boiling water until float.
4. Scoop out, strain and coat with soy bean powder.

入廚貼士 | Cooking Tips

- 即煮即食，趁熱享用。
- Serve while it is hot.

20 粒
20 pcs

30~35 分鐘
30~35 minutes

糯米糍
Coconut Glutinous Rice Balls

粉糰 | Dough

糯米粉 200 克
糖 70 克
椰汁 110 毫升
水 80 毫升

200g glutinous rice flour
70g sugar
110ml coconut milk
80ml water

餡料 | Filling

花生 50 克
芝麻 2 湯匙
糖 1 湯匙

50g peanuts
2 tbsps sesame seeds
1 tbsp castor sugar

飾面 | Garnish

日本片栗粉 / 椰絲適量

Some Japanese corn flour/ desiccated coconut

做法 | Method

1. 糯米粉置碗中，加入椰汁、糖和水緩緩拌勻，直至完全沒有粉粒。
2. 花生切碎，加入芝麻和砂糖拌勻，即成餡料。
3. 在正方形的糕盤上掃少許油，倒入粉漿，隔水蒸 30 分鐘。取出，放置 10 分鐘，稍涼。
4. 切割成約 20 份，在中央捏成凹穴，包少許餡料，收口。沾上椰絲或片栗粉即成。

1. Place glutinous flour in a mixing bowl, add coconut milk, sugar and water. Stir thoroughly until well blended.
2. Chop peanuts finely, add sesame seeds and castor sugar, mix well as filling.
3. Brush some oil on a square mould. Pour in batter and steam for 30 minutes, take out and leave for 10 minutes until cool.
4. Cut the dough into 20 portions. Knead a hollow in the centre of each piece, wrap in some filling and seal well. Coat with some desiccated coconut or Japanese corn flour, serve.

朱古力糯米糍
Chocolate Glutinous Rice Balls

20 粒
20 pcs

20 分鐘
20 minutes

粉糰 | Dough

牛奶 160 毫升
糯米粉 110 克
糖 55 克
可可粉 10 克

160ml milk
110g glutinous rice flour
55g sugar
10g cocoa powder

朱古力餡 | Chocolate Filling

牛奶 220 克	220g milk
糖 50 克	50g sugar
牛油 35 克	35g butter
麵粉 15 克	15g flour
粟粉 15 克	15g cornstarch
可可粉 15 克	15g cocoa powder
蛋黃 3 隻	3 egg yolks

做法 | Method

1. 糯米粉與可可粉篩勻後同置碗中，加入糖和牛奶徐徐拌勻，直至沒粉粒。倒入已掃油的糕盤，隔水蒸 20 分鐘，取出，置放 10 分鐘，稍涼後分切成約 20 份。
2. 蛋黃與糖置碗中打至淡黃色；麵粉、粟粉和可可粉篩勻，加入 50 毫升牛奶調勻。
3. 用餘下的牛奶煮熱，撞入已打散的蛋黃中，拌勻，再加入牛奶粉糊中，一邊煮一邊拌至濃稠。離火，加入牛油搓勻。
4. 待涼後，包入粉糰中，篩上可可粉即可。

1. Sift glutinous rice flour and cocoa powder in a mixing bowl. Add sugar and milk gradually and stir well until fully blended. Pour mixture into a greased mould and steam for 20 minutes, take out and leave for 10 minutes until cool. Cut the dough into 20 portions.
2. Beat egg yolks and sugar in a mixing bowl until light yellow. Sift flour, cornstarch and cocoa powder, add 50ml milk and stir well.
3. Heat remaining milk, pour into egg yolk mixture and stir well. Then pour mixture into milk paste, stir over heat until thick. Remove from heat, add butter and knead thoroughly.
4. Leave chocolate filling to cool down. Wrap some filling with a piece of dough and roll it into a ball. Garnish with cocoa powder and serve.

3~4 人
Serves 3~4

30 分鐘
30 minutes

豆蓉拌丸子

Green Beans with Glutinous Balls

材料 | Ingredients

開邊綠豆 100 克	100g shelled green beans
糖 40 克	40g sugar
小丸子（煮熟）適量	Some cooked glutinous balls
椰果適量	Some coconut milk
淡奶適量	Some evaporated milk

做法 | Method

1. 開邊綠豆用清水浸 1 小時，瀝乾水分。
2. 燒熱鍋，開邊綠豆置煲中，水略浸過豆面，加入糖，不停攪拌至水開始收乾，熄火。
3. 待涼後，放入雪櫃中冷藏。
4. 食用時，以椰果和淡奶拌食。

1. Soak green beans in water for an hour and strain.
2. Heat a wok, put green beans and fill in water until it covers the beans. Add sugar and stir continuously while cooking until water starts to reduce. Turn off heat.
3. Leave it to cool and put into a refrigerator.
4. Serve with coconut milk and evaporated milk.

入廚貼士 | Cooking Tips

- 小丸子在南貨店有售，或用糯米粉與水搓成粉糰，再分成小粒，煮熟亦可。
- Ready-to-use glutinous balls can be bought in Shanghaiese grocery stores. It can be made by yourself: mix glutinous rice flour with water, knead into a dough, divide into little balls and then cook in boiling water.

酒釀小丸子

Glutinous Rice Balls in Wine

4~6 人
Serves 4~6

20~25 分鐘
20~25 minutes

材料 | Ingredients

糯米粉 100 克	100g glutinous rice flour
糖 20 克	20g sugar
酒釀 3 湯匙	3 tbsps fermented glutinous rice
水 1 杯	1 cup water
熱水 1/2 杯	1/2 cup hot water

做法 | Method

1. 碗中放糯米粉，慢慢加入熱水，搓揉至粉與水混合。再揉成軟度適中的麵糰，然後揉搓成球狀。
2. 燒熱一鍋水，放入小丸子煮至浮起。
3. 撈出小丸子，放入冷水中，待冷卻後，瀝乾水分。
4. 燒熱鍋，加入水與糖煮滾，待涼後加入酒釀和小丸子，拌勻即可。

1. Put glutinous rice flour into a bowl, add hot water slowly, mix well and knead into a dough. Then divide into rice balls.
2. Put rice balls into boiling water and cook until float.
3. Scoop out rice balls and place into cold water, leave to cool and strain.
4. Heat a pot, boil water and sugar together, leave to cool. Add fermented glutinous rice and rice balls and mix well. Ready to serve.

4~6 人
Serves 4~6

10 分鐘
10 minutes

芒果班戟
Mango Pancake

材料 | Ingredients

高筋麵粉 50 克	50g high gluten flour
糖 20 克	20g sugar
雞蛋 2 隻	2 eggs
牛油溶液 8 克	8g melted butter
芒果適量	Some mangoes
甜忌廉（已拂打）適量	Some sweetened cream (whipped)
牛奶 100 毫升	100ml milk
淡忌廉 10 毫升	10ml whipping cream
水 25 毫升	25ml water

做法 | Method

1. 雞蛋置碗中，與糖拂打至淡黃色。
2. 牛奶、淡忌廉和水分次加入雞液中，充分拌勻。將已篩好的高筋麵粉拌入蛋糊中，加入牛油溶液拌勻。
3. 燒熱平底鍋，薄薄塗上一層牛油，舀一杓麵糊於鍋中，煎至兩面金黃色，待涼。
4. 將忌廉和芒果粒加入班戟內，包好即可享用。

1. Place eggs into a bowl and whisk with sugar until light yellow.
2. Add milk, whipping cream and water into egg liquid one by one until blended. Fold in sifted high gluten flour, add melted butter and stir well.
3. Heat a pan, brush a thin layer of butter, pour a ladle of batter and pan-fry until both sides become brownish, leave to cool.
4. Add cream and mango dice onto pancake, wrap and serve.

士多啤梨熱香餅

Strawberry Hotcakes

熱香餅 | Hotcake

雞蛋 1 隻
麵粉 65 克
牛油 40 克
糖 20 克
泡打粉（發粉）2 茶匙
牛奶 25 毫升

1 egg
65g flour
40g butter
20g sugar
2 tsps baking powder
25ml milk

3~4 人
Serves 3~4

35~50 分鐘
35~50 minutes

士多啤梨汁 | Strawberry sauce

糖 1 湯匙
士多啤梨醬 1 茶匙
粟粉 1/2 茶匙
士多啤梨蓉適量
水 4 湯匙
1 tbsp sugar
1 tsp strawberry jam
1/2 tsp corn flour
Some strawberry puree
4 tbsps water

做法 | Method

1. 牛油放碗中，隔水座溶。
2. 雞蛋放碗中打散，加入糖攪至淡黃色，再加入奶和牛油溶液，充份打勻。
3. 加入已篩勻的麵粉和泡打粉，拌勻成粉糊。
4. 燒熱平底鍋，抹上一層薄薄牛油，舀一杓麵糊於鍋中，煎至兩面金黃色。
5. 燒熱鍋，加入糖、粟粉和水同煮滾，加入士多啤梨醬拌勻，熄火。再加入士多啤梨蓉拌勻，即可與熱香餅一同享用。

1. Place butter in a bowl and melt above hot water.
2. Beat egg in a bowl, add sugar and stir until light yellow. Add milk and melted butter and whisk thoroughly.
3. Sift flour and baking powder together, add mixture into egg liquid and stir into a batter.
4. Heat a pan and brush a thin layer of butter, put a ladle of batter and pan-fry until both sides turn brownish.
5. Heat a wok, add sugar, corn flour and water together and bring to a boil. Add strawberry jam, mix well and turn off heat. Add strawberry purée and mix well. Serve with hotcakes.

10 粒
10 pcs

20~25 分鐘
20~25 minutes

大福 Daifuku

材料 | Ingredients

糯米粉 100 克
糖 70 克
鹽少許
片栗粉適量（撲手用）
水 125 毫升
水 2 湯匙（調節粉糰的軟硬度）

100g glutinous rice flour
70g sugar
A pinch of salt
Some Japanese corn flour
(to make hands non-sticky)
125ml water
2 tbsps water
(to adjust the texture of dough)

餡料 | Filling

紅豆粗粒蓉 150 克

150g red bean chunky purée

做法 | Method

1. 糯米粉、水和糖放碗中拌勻，倒入不黏底糕盤隔水蒸 15 分鐘；紅豆餡料分成 10 份。
2. 將已蒸熟的粉糰移入鍋中，用小火加熱，不停攪動至幼滑且濃稠，慢慢加水至軟硬適中，拌勻至黏稠（應該更具透明感），即可。
3. 工作枱撒上生粉，把熟粉糰放上，待稍涼後揉成細長條狀，分成 10 份。
4. 把小粉糰放在掌心上，壓扁，包上餡料，即成。

1. Mix glutinous rice flour, sugar and water in a bowl. Pour mixture into a non-sticky mould and steam for 15 minutes. Divide red bean filling into 10 portions.
2. Place steamed dough into a pot, cook over low heat, keep stirring until thick and smooth. Add water gradually until an optimal texture is attained. Stir until mixture becomes translucent and set aside.
3. Sprinkle Japanese corn flour onto a workbench, place the dough to cool down. Knead it into a long and thin rod, then divide into 10 portions.
4. Place a piece of dough on a palm, press to flatten, wrap in some filling. Serve.

豆沙餅

Red Bean Pancake

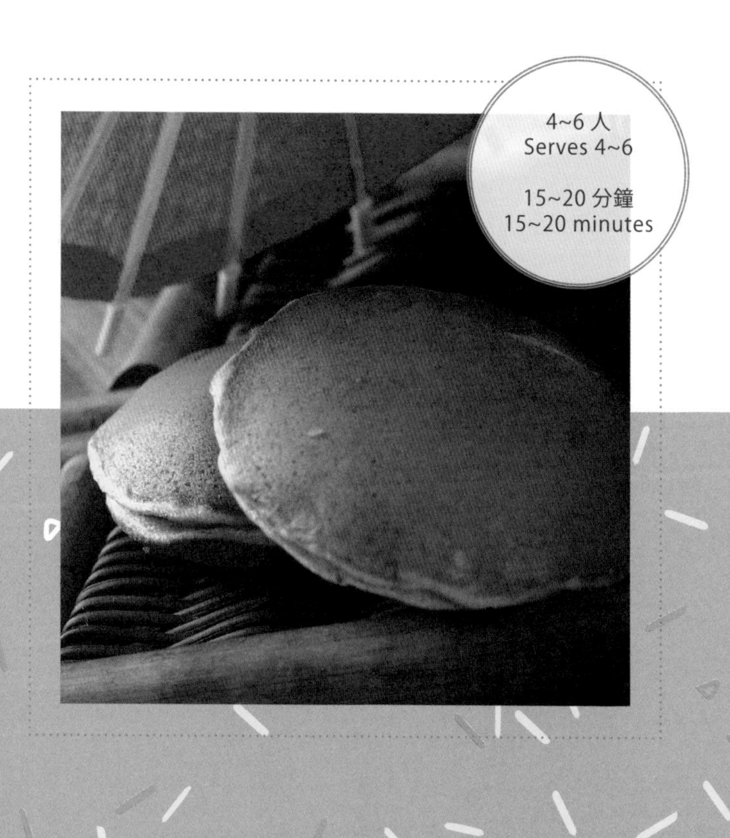

4~6 人
Serves 4~6

15~20 分鐘
15~20 minutes

材料 | Ingredients

雞蛋 3 隻
低筋麵粉 180 克
糖 90 克
蜂蜜 3 湯匙
紅豆蓉適量
泡打粉（發粉）1 茶匙
水 125 毫升
3 eggs
180g low gluten flour
90g sugar
3 tbsps honey
Some red bean purée
1 tsp baking powder
125ml water

做法 | Method

1. 雞蛋放碗中打散，加入糖充分拂打至淡黃色，加入蜂蜜和水拌勻。
2. 低筋麵粉和發粉同篩勻，加入（1）中拌勻。
3. 燒熱平底鍋，塗上一層薄薄牛油，舀一杓麵糊於鍋中，煎至麵糊起氣泡，翻轉，煎至金黃。
4. 把紅豆蓉放在一片煎餅上，再蓋上另一片，即可。

1. Beat eggs in a mixing bowl, add sugar and whisk until light yellow. Add honey and water and stir thoroughly.
2. Sift low gluten flour and baking powder together, add in the egg mixture from step (1) and mix well.
3. Heat a pan and brush a thin layer of butter. Pour a ladle of batter and pan-fry until air bubbles appear, then flip it over and pan-fry until both sides turn brownish.
4. Place some red bean purée on a pancake and stack another pancake on top. Serve.

3~4 人
Serves 3~4

20~25 分鐘
20~25 minutes

砵仔糕

Rice Pudding in Bowl

材料 | Ingredients

粘米粉 100 克	100g rice flour
片糖 80 克	80g rock sugar
澄麵 60 克	60g tang starch
糯米粉 1 湯匙	1 tbsp glutinous rice flour
即食紅豆適量	Some ready-to-eat red beans
水 500 毫升	500ml water

做法 | Method

1. 粘米粉、糯米粉和澄麵放碗中同篩勻，用 150 毫升水拌勻，調成麵糊。
2. 燒熱鍋，加入糖和 350 毫升水同煮滾，趁熱撞入麵糊中，不停用木杓拌勻成幼滑麵漿。
3. 將麵漿倒入已掃油之器皿上，加入紅豆，用大水隔水蒸 20 分鐘至熟，即可享用。

1. Sift rice flour, glutinous rice flour and tang starch in a bowl together, mix with 150 ml of water and stir into a smooth batter.
2. Heat a wok and boil sugar and water together, pour into batter while it is hot. Stir with a wooden ladle continuously until blended.
3. Pour batter into a greased container and add red beans. Steam over high heat for 20 minutes until cooked. Serve.

入廚貼士 | Cooking Tips

- 器皿的尺碼大小會影響烹調時間，宜自行調校烹調時間。
- Since the size of the mould will directly affect cooking time required, you may adjust the time accordingly.

葉香茶粿

Glutinous Rice Cake in Banana Leaf

粉糰 | Dough

糯米粉 80 克
粘米粉 80 克
糖 15 克
蕉葉（大）2 片
鹽少許
水 160 毫升

80g glutinous rice flour
80g rice flour
15g sugar
2 banana leaves (large)
Pinch of salt
160ml water

8~10 個
8~10 pcs

30~35 分鐘
30~35 minutes

餡料 | Filling

冬菇 2 朵
甜菜脯 10 克
蝦米 10 克
花生碎 10 克

2 dried black mushrooms
10g sweet preserved turnip
10g dried shrimps
10g chopped peanuts

調味料 | Seasonings

糖 1/2 茶匙
生粉 1/2 茶匙
鹽 1/4 茶匙
水 1 湯匙

1/2 sugar
1/2 tsp cornstarch
1/4 tsp salt
1 tbsp water

做法 | Method

1. 糯米粉、粘米粉和鹽放碗中同篩勻。
2. 燒熱鍋，加入糖和水煮滾，撞入糯米粉和粘米粉中，搓成粉糰，再用布蓋上備用。
3. 餡料洗淨，切碎。燒熱鍋，加入餡料和調味料炒熟。
4. 粉糰分成 8~10 份，將粉糰搓長少許，然後包上餡料，收口後，用蕉葉包好。以大火隔水蒸 10~15 分鐘，即成。

1. Sift glutinous rice flour, rice flour and salt in a bowl together.
2. Heat a wok, add sugar and water together and bring to a boil. Pour into flour and mix well, knead into a dough, cover with a cloth and set aside.
3. Rinse filling ingredients, chop up finely, add seasoning. Heat a wok and stir-fry until done.
4. Divide dough into 8~10 portions. Roll each portion into an oblong shape, wrap in some filling, seal well. Wrap with a piece of banana leaf. Steam over high heat for 10~15 minutes. Serve.

4~6 人
Serves 4~6

15~20 分鐘
15~20 minutes

白芝麻卷

White Sesame Seed Roll

材料 | Ingredients

冰糖 200 克	200g rock sugar
白芝麻 80 克	80g white sesame seeds
馬蹄粉 80 克	80g water chestnut powder
豬油 / 菜油 1/2 湯匙	1/2 tbsp lard / vegetable oil
清水 800 毫升	800ml water

做法 | Method

1. 白芝麻洗淨，瀝去水分。燒熱鍋，放白鑊中炒香，放涼，加入 400 毫升水，用攪拌器磨碎，再用布袋濾過。
2. 燒熱鍋，用 300 毫升水把冰糖煮融。馬蹄粉置碗中，加入 100 毫升水拌勻成粉漿。
3. 燒熱鍋，將已過濾的白芝麻水和豬油加入已煮滾的糖水中，煮 5~10 分鐘，熄火。
4. 待涼 5 分鐘，加入馬蹄粉漿，拌勻。
5. 糕盤掃油，倒入兩湯杓芝麻粉漿，用猛火蒸 5 分鐘。取出稍涼後，捲成一長條，再切成 3~4 件，上碟。

1. Rinse sesame seeds, strain excess water. Heat a wok and stir-fry sesame seeds without adding oil until aromatic. Then place in a mixing bowl, add 400ml of water and grind with a mixer. Then strain with a cotton bag.
2. Heat a wok, boil rock sugar with 300ml of water until dissolved. Mix 100ml of water with water chestnuts powder into a batter.
3. Heat a wok, add filtered sesame seed water and lard into boiling syrup, cook for 5~10 minutes, turn off heat.
4. Leave for 5 minutes, add water chestnut batter and mix thoroughly.
5. Brush some oil on a cake mould. Pour in 2 ladles of sesame seed batter and steam over high heat for 5 minutes. Take out and leave to cool. Roll into a long rod, cut into 3~4 pieces and serve.

巧手糕點

編著
方芍堯

編輯
紫彤　陳芷欣

美術設計
Venus Lo

排版
葉青

翻譯
Tracy

攝影
Wilson Wong

出版者
萬里機構出版有限公司
香港鰂魚涌英皇道1065號東達中心1305室
電話：2564 7511
傳真：2565 5539
電郵：info@wanlibk.com
網址：http://www.wanlibk.com
http://www.facebook.com/wanlibk

萬里機構

萬里 Facebook

發行者
香港聯合書刊物流有限公司
香港新界大埔汀麗路36號
中華商務印刷大廈3字樓
電話：2150 2100
傳真：2407 3062
電郵：info@suplogistics.com.hk

承印者
美雅印刷製本有限公司

出版日期
二零一八年十月第一次印刷

Published in Hong Kong.
ISBN 978-962-14-6866-6